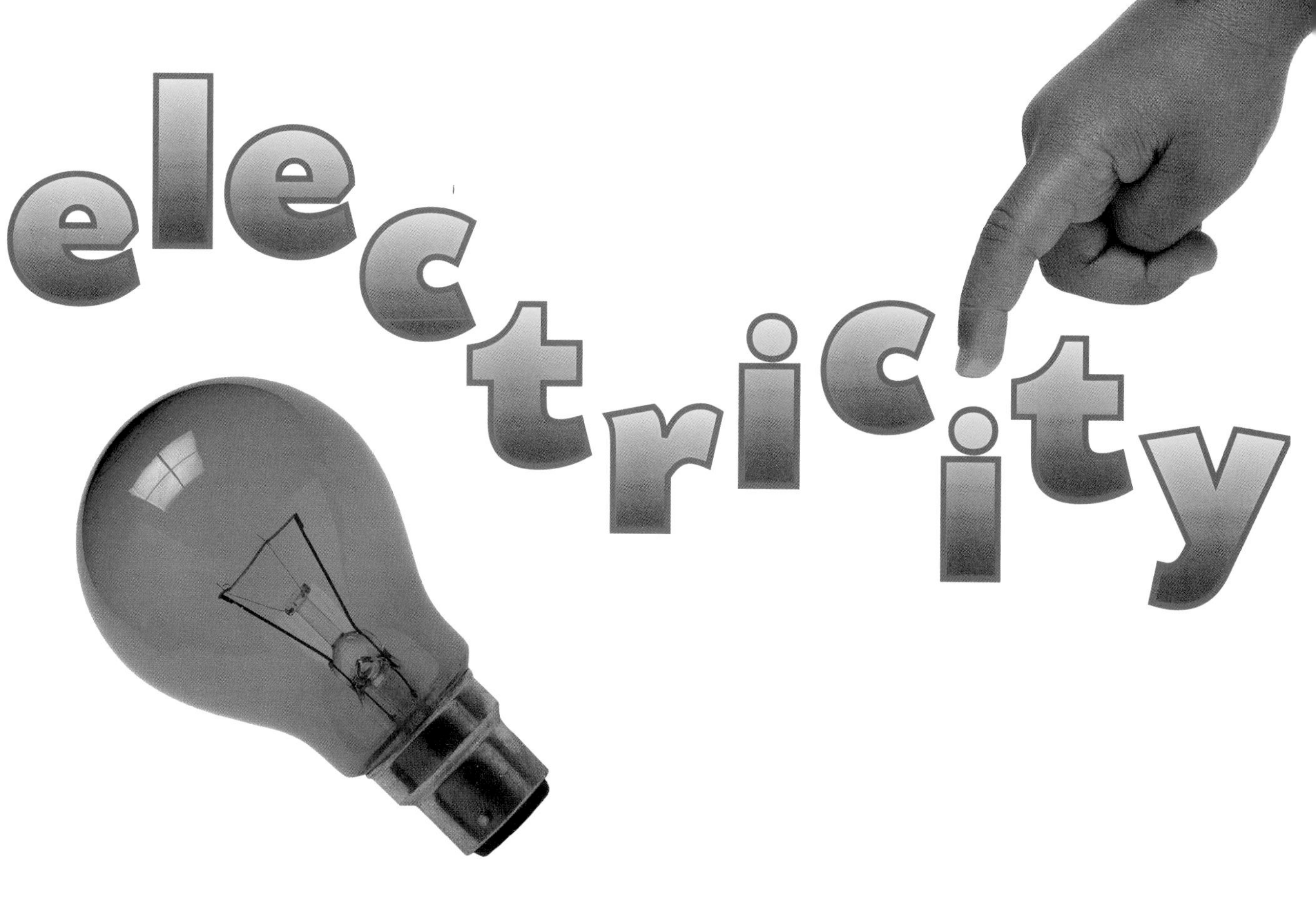

Terry Jennings

BBC Education
201 Wood Lane
London W12 7TS

ISBN 0 563 37334 2

Editor: Christina Digby
Designer: Jo Digby
Picture research: Helen Taylor
Educational advisers: Su Hurrell, Samina Miller, Shelagh Scarborough
Photographers: John Jefford, Simon Pugh (photographs of children)
Illustrator: Philip Dobree

With grateful thanks to:
Rebecca Digby, Peter Kofi Arboine-Joseph, Pascalle Matherson-Frederick, The Wind Energy Group

Researched photographs ©: Bruce Coleman Ltd pp. 18 (top), 20 and 21; Robert Harding Picture Library Ltd p. 22; Image Bank p. 13; Milepost 92 1/2 p. 11; National Power Picture Unit pp. 16 and 17

Printed in Belgium by Proost
Colour origination by Goodfellow & Egan, Peterborough

Contents

How does a torch work? page 5

How does a battery work? page 7

What is an electrical circuit? page 9

What do we use electricity for? page 11

How does electricity give us light? page 13

How does electricity give us heat? page 15

How is electricity made? page 17

How does electricity get to our homes? page 19

Can we get electricity from the wind? page 21

Why should we save electricity? page 23

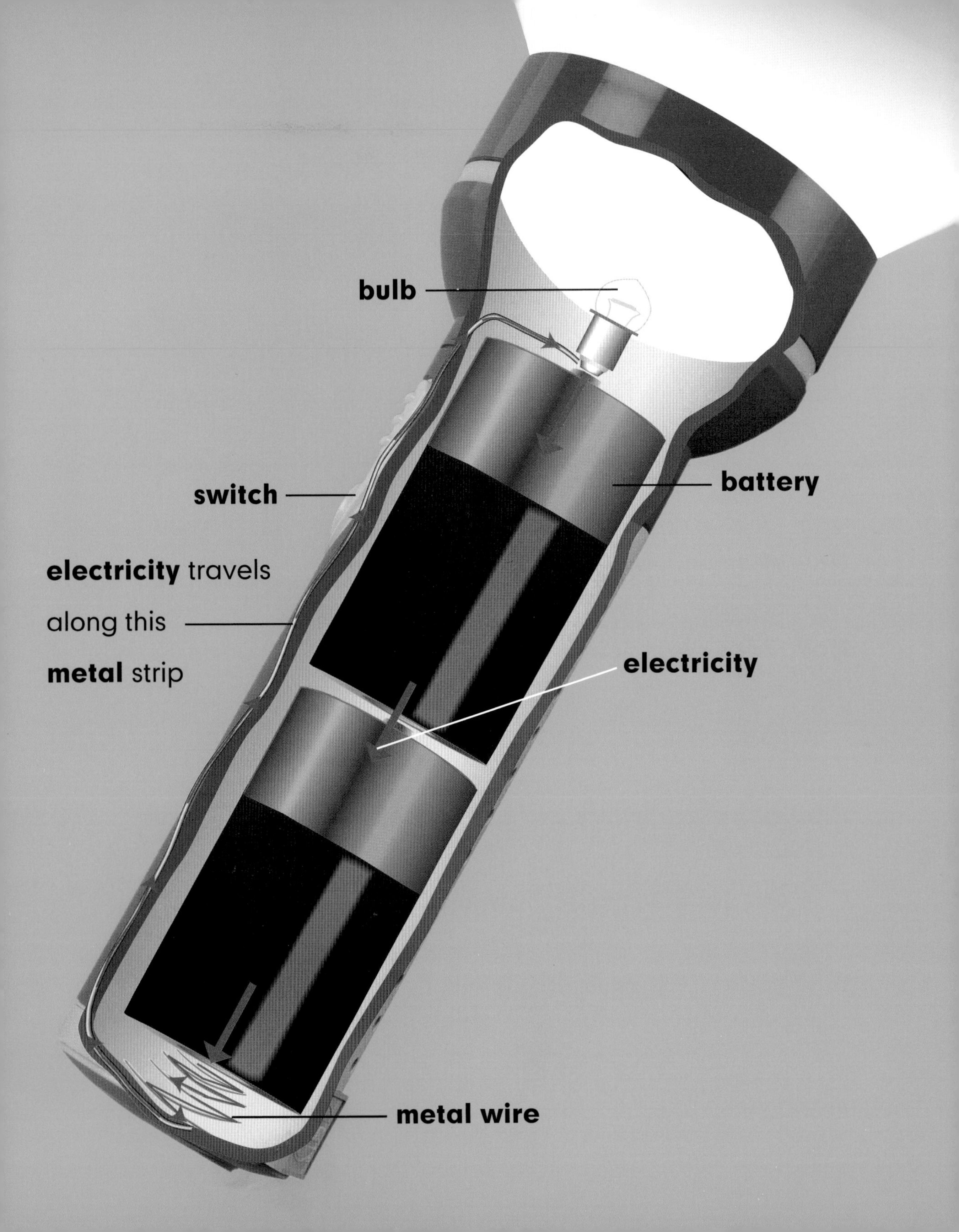
bulb
switch
battery
electricity travels
along this
metal strip
electricity
metal wire

How does a torch work?

A torch is a useful kind of light. You can carry it around. It uses electricity to make light. A torch gets its electricity from batteries. Most torches have two batteries. All torches have a small light bulb and a switch.

When you press the switch of a torch, electricity flows from the batteries. The electricity flows along thin strips of metal to the bulb. It makes the bulb light up. Then the electricity goes back into the batteries.

Take a torch apart to see what it is like inside. Be careful not to lose any of the parts.

This toy car uses **electricity** from a **battery**.

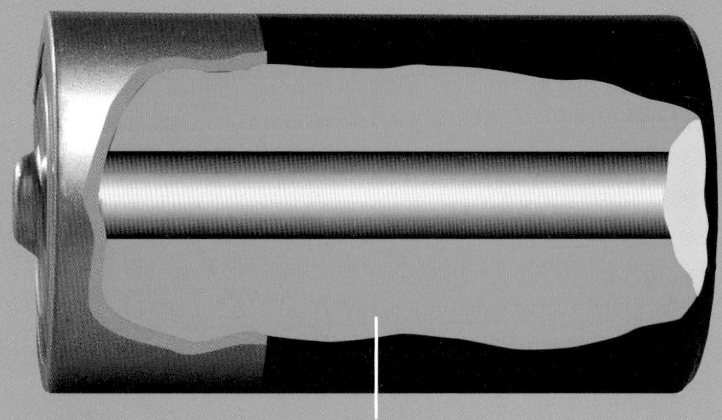

The **chemicals** inside a **battery** make **electricity**.
They are dangerous if they leak out.

How does a battery work?

A torch uses small amounts of electricity. Some toys and small machines also need small amounts of electricity to make them work. They get their electricity from batteries.

Inside a battery there are special chemicals. These chemicals work together to make electricity. The electricity flows through wires to make the toy or machine work. When the chemicals are used up the battery stops working.

Batteries come in all shapes and sizes.

Do any of your toys use batteries?

Ask an adult to cut the covering from the ends of two pieces of wire.

Join one end of each wire to a torch bulb in a holder.

Fix a paperclip to the other end of each wire.

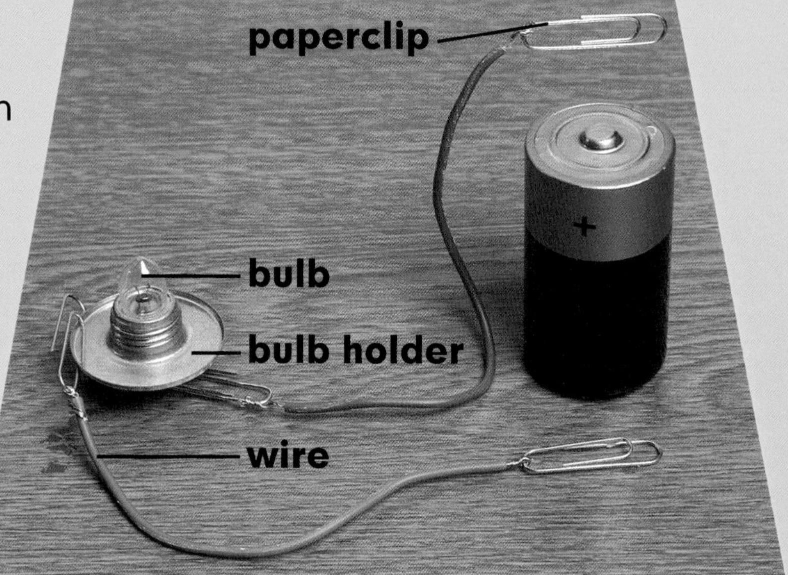

Tape the paperclips to the battery.

See how the bulb lights up.

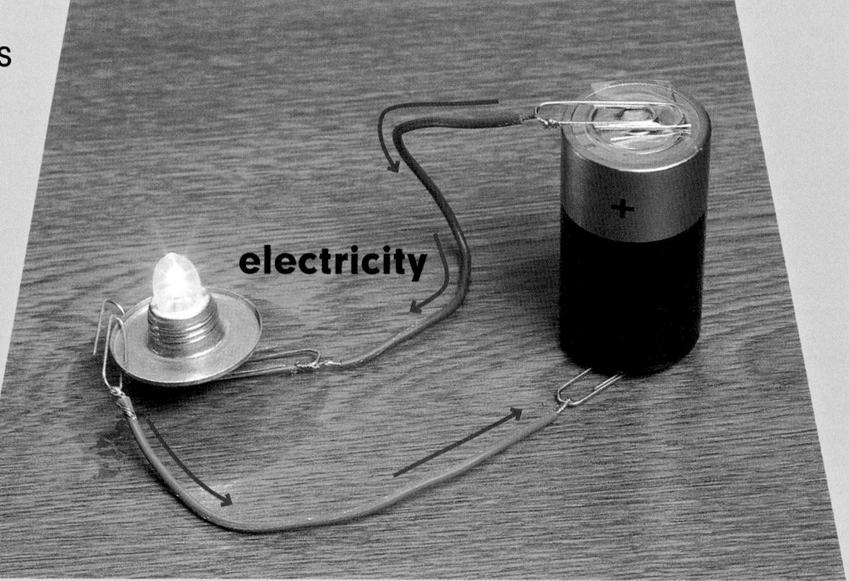

Undo one of the wires. What happens?

What is an electrical circuit?

To make electricity come from a battery, you need to give it a path to flow along. This path is called a circuit. Most circuits are made of wires. A circuit goes in a loop.

Electricity flows around the circuit – through the wires, the bulb and the battery. Then it goes back into the battery. Electricity cannot flow if there is a gap in the circuit.

A switch is like a little gate. When the switch is 'off' it makes a gap in the circuit. Then the electricity cannot flow and the bulb will not light. When you press the switch 'on' the gap closes again. Now the electricity can flow, and the bulb lights.

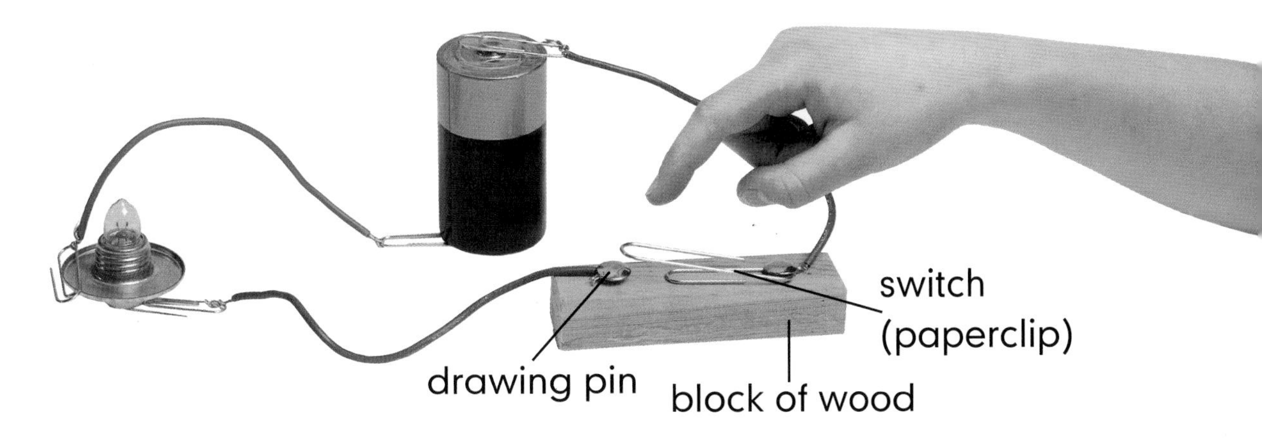

How many **electrical** things can you find in your home?

How many **electrical** things have you used today?

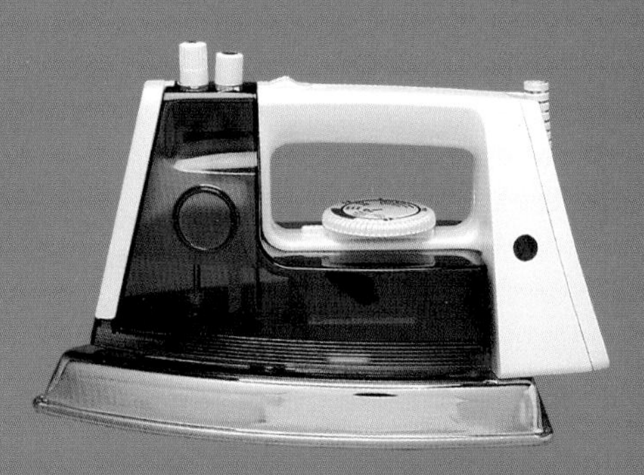

What do we use electricity for?

We cannot see electricity, but we can see where it is working. Our lives would be very different without electricity.

Electricity lights our homes and schools and shops. It also lights our streets. Many people use electricity for cooking and heating. Fridges use electricity to keep our food cool. Electricity also works our televisions and computers. Trams and some trains use electricity to make them move. All of these things use a lot of electricity. They need too much electricity to be able to get it from batteries.

This train is moved by **electricity**. It gets the electricity from overhead wires.

How many **electric light bulbs** can you find in your home?

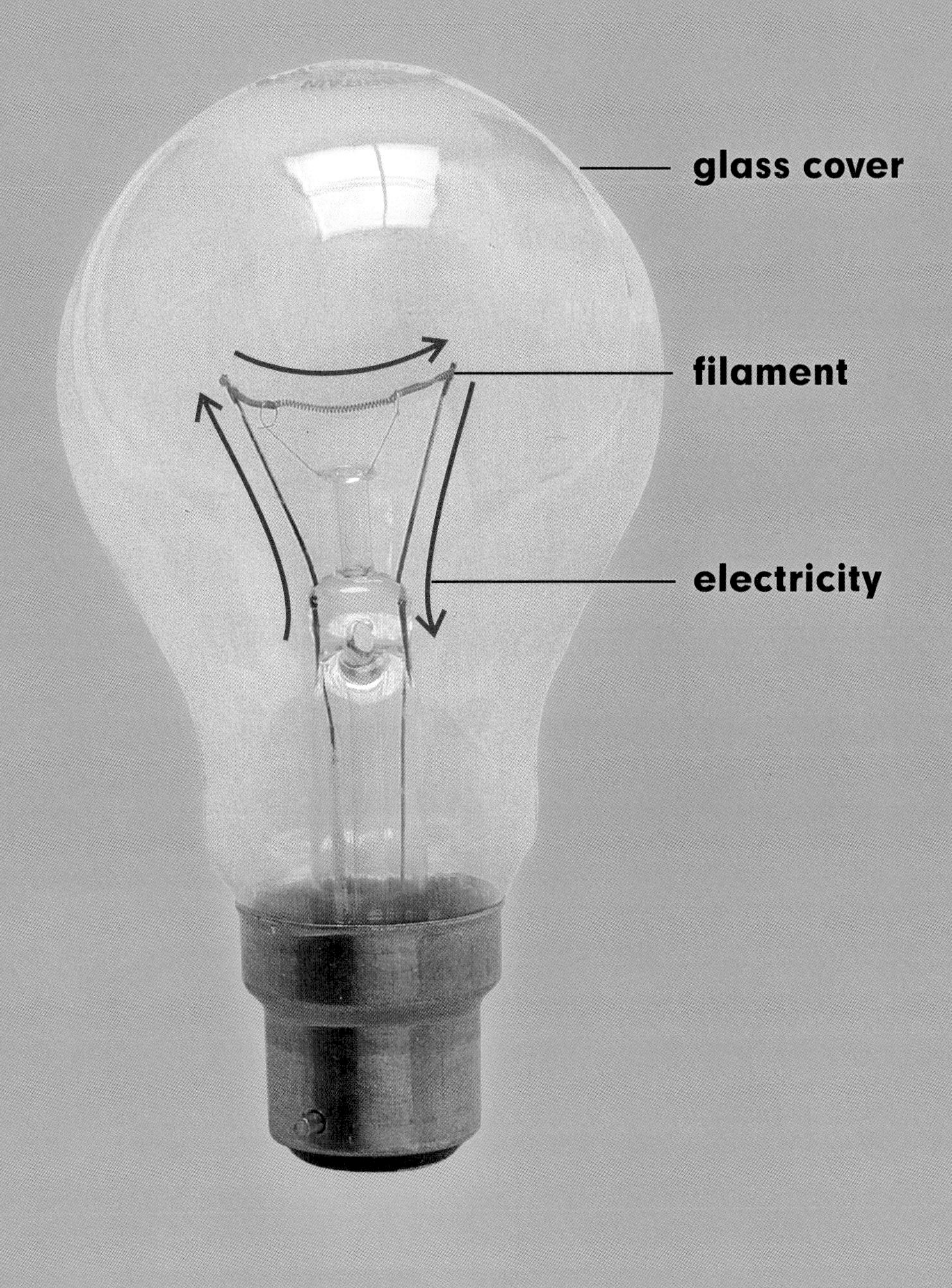

How does electricity give us light?

Electric light bulbs give us light. When you turn on a light switch, electricity goes through the wires. The electricity goes into the bulb. Inside the bulb is a very thin coil of wire. This is called the filament.

Electricity goes through the filament and makes it heat up. The filament gets hotter and hotter. Soon it glows white hot. The white hot filament gives out the light you see.

Sometimes we get light from special tubes. These tubes also use electricity.

Have you seen **light tubes** like these?

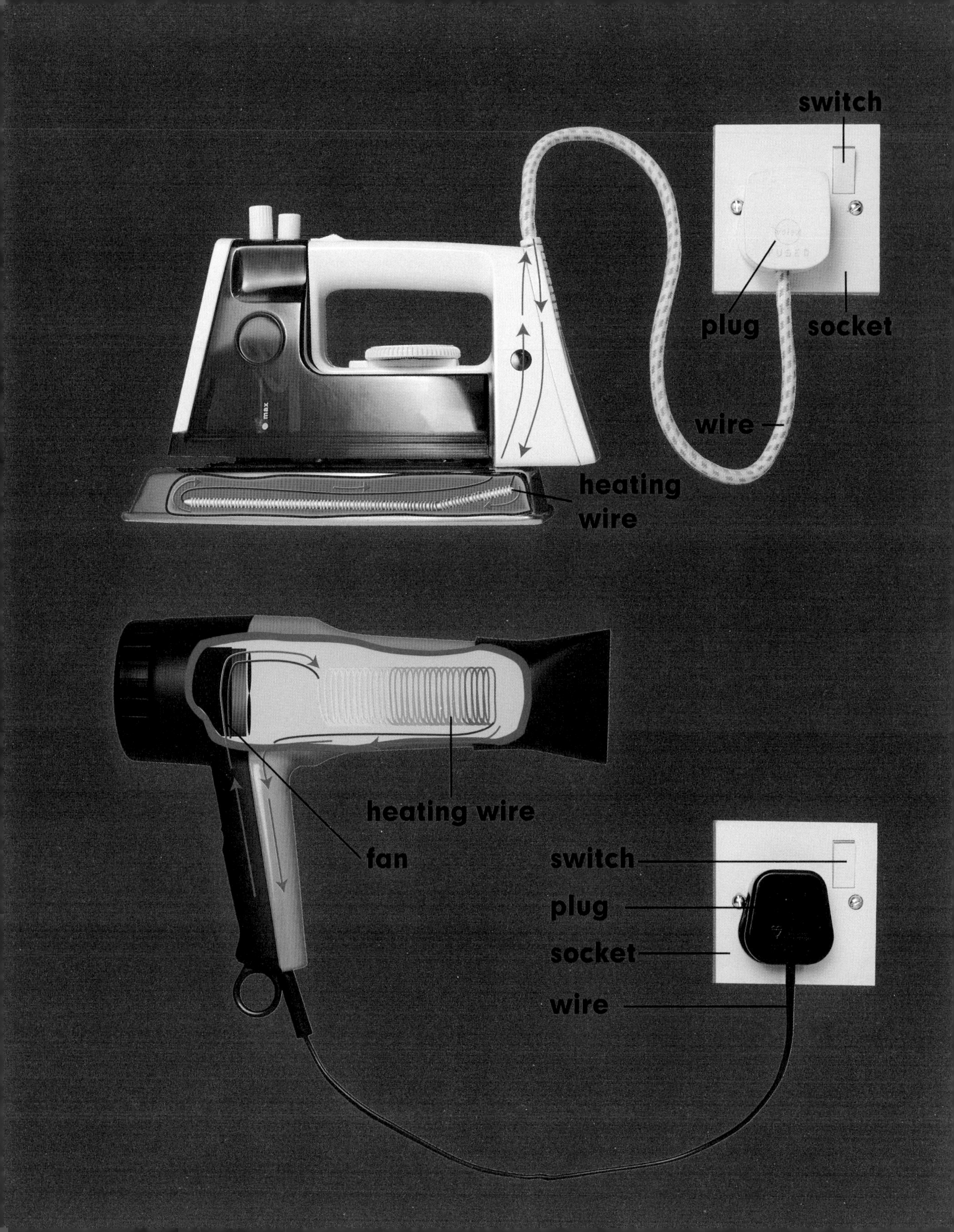
switch
plug
socket
wire
heating
wire
heating wire
fan
switch
plug
socket
wire

How does electricity give us heat?

An iron has special heating wires inside it. These wires get very hot when electricity flows through them. As the wires get hot, the iron heats up.

A hair dryer has small heating wires inside it. A fan blows air over the hot wires. The air heats up so that you can dry your hair with it.

There are also heating wires inside most electric cookers.

The **heating wires** inside a toaster go **red** when they get hot. They will **burn** you if you touch them.

This is one of the machines which makes **electricity** inside a **power station**.

How is electricity made?

Nearly all the electricity we use is made in a large building called a power station. Inside a power station are big machines called generators. The generators make electricity.

Most power stations burn coal, oil or gas. When these fuels burn they give out heat. The heat is used to turn water into steam. Steam can push very hard. Have you seen a lid being pushed up by steam?

The steam in a power station rushes along big pipes. The steam pushes the generators round very fast. The generators make electricity.

This **power station** burns oil. Is there a power station near you?

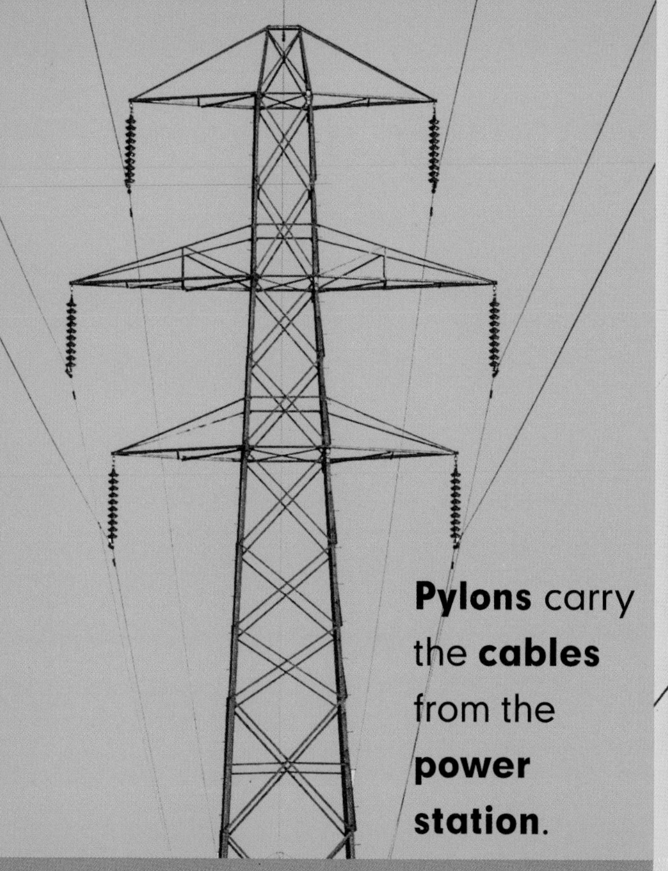

Pylons carry the **cables** from the **power station**.

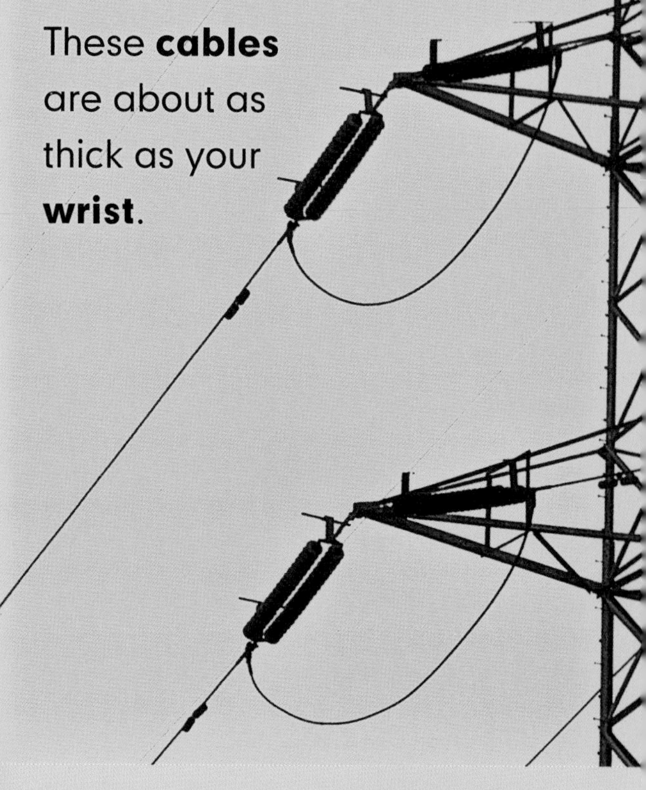

These **cables** are about as thick as your **wrist**.

Thinner **cables** bring the electricity into your home.

Never play with plugs and sockets.

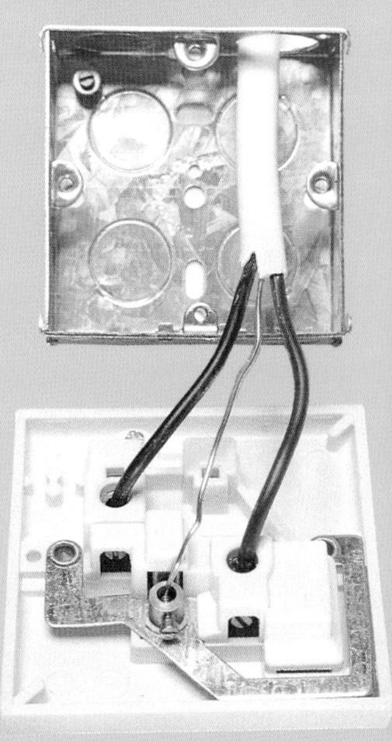

We use electricity from **sockets** in the walls.

How does electricity get to our homes?

Electricity goes from the power station to homes, shops and factories. It travels along thick wires called cables. Metal towers called pylons hold the cables high above the ground.

In towns and cities the cables go under the ground. Thinner cables bring the electricity into your home. The wires that carry the electricity round your home are inside the walls and ceilings or under the floors for safety.

We use electricity from sockets in the walls. The plug on an electric lamp or machine fits into a socket. When it is switched on, the electricity goes along the wire to the lamp or machine.

Electricity from power stations can be very dangerous.
Never play with plugs and sockets.

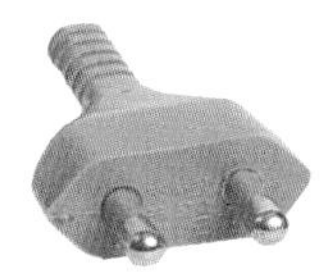

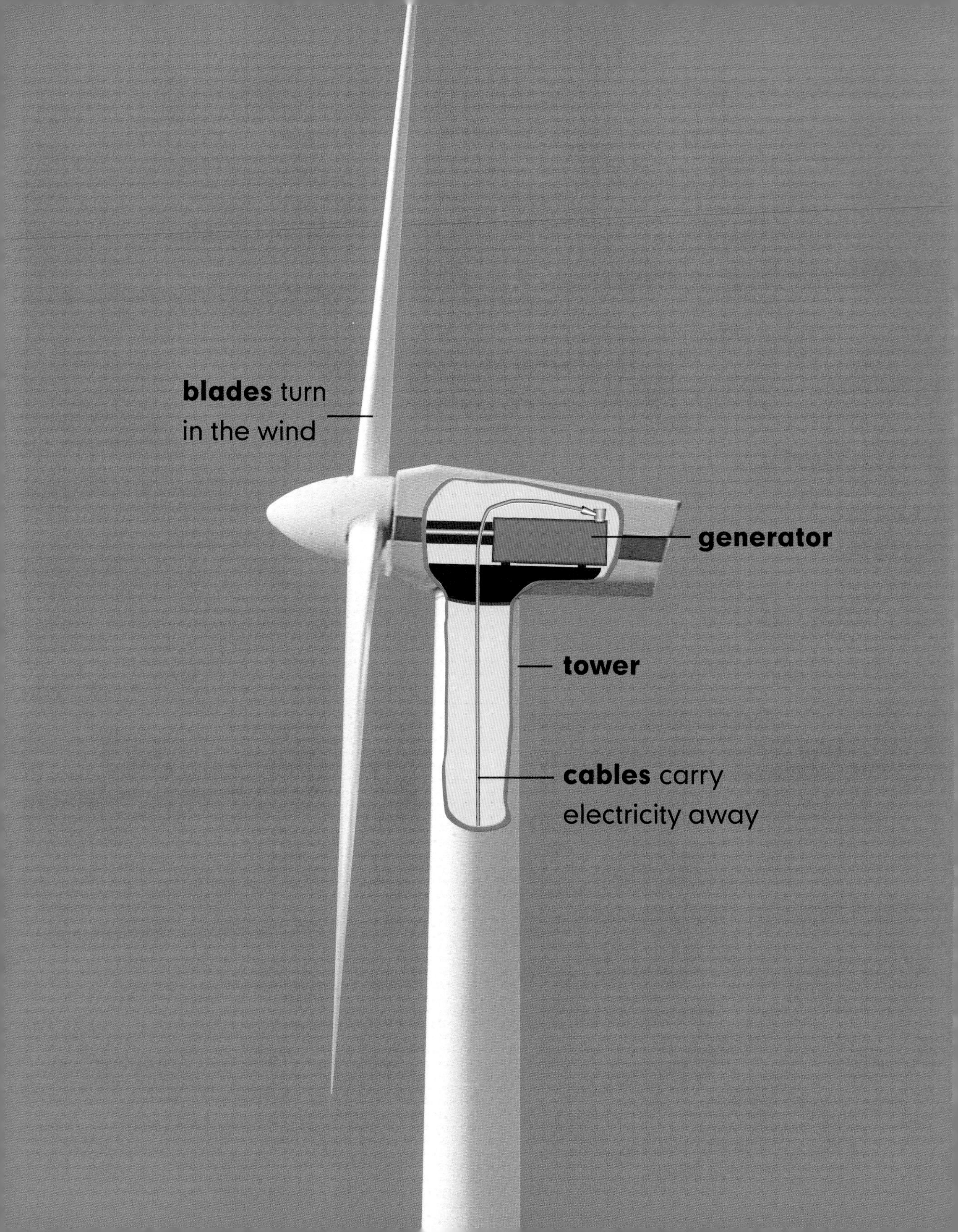

blades turn
in the wind
generator
tower
cables carry
electricity away

Can we get electricity from the wind?

Most power stations burn coal, oil or gas. But one day we will have used up all the coal, oil and gas.

The wind can also be used to make electricity. This is done by windmills with huge blades. When the wind blows hard it pushes the blades to turn them round. As the blades turn, they turn a generator to make electricity.

Unlike coal, oil and gas, the wind will never run out. The problem is that it is not always windy. Windmills do not dirty the air with smoke like many power stations. However, they are noisy.

A **windfarm** is a collection of **windmills**.

Oil rigs drill through the sea floor to get the **oil** or **gas**.

drill from rig

bottom of the sea

oil

Why should we save electricity?

Most power stations burn coal, oil or gas to make electricity. We get most of our oil and gas from deep under the sea. Some of the oil is spilled. It dirties, or pollutes, the sea and kills sea birds and fish.

When coal, oil and gas are burned, they pollute the air with smoke and gases. Sometimes the smoke and gases mix with rain. This makes the rain acid like vinegar. Acid rain damages buildings. It can kill trees and fish.

The only way we can save electricity is to use less of it. That way there will be less pollution. Save electricity by . . .

not filling the kettle too full

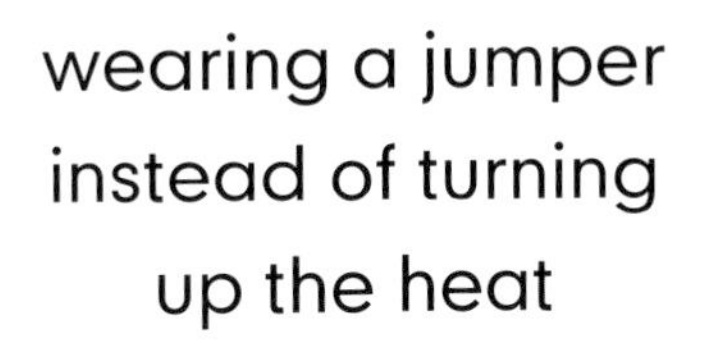
wearing a jumper instead of turning up the heat

using light bulbs like this

Index

battery 5, 7, 9, 11

bulb 5, 9, 13, 23

cables 19

generator 17, 21

heat 11, 15, 17, 23

lights 11, 13

plugs 19

pollution 23

power station 17, 19

pylon 19

sockets 19

switch 5, 9, 13

torch 5, 7

windmills 21

wires 7, 9, 11, 13, 15, 19

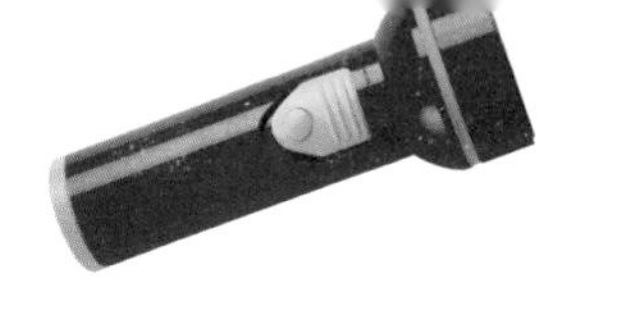

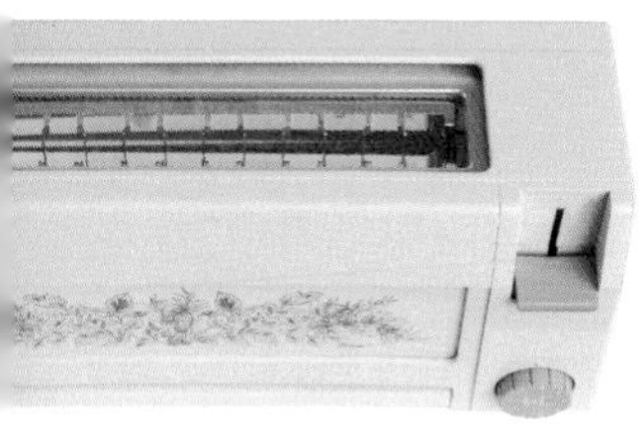

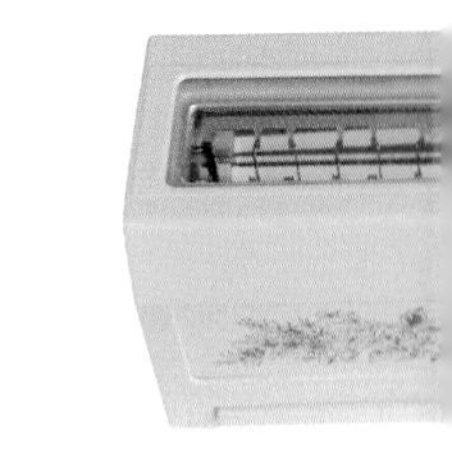

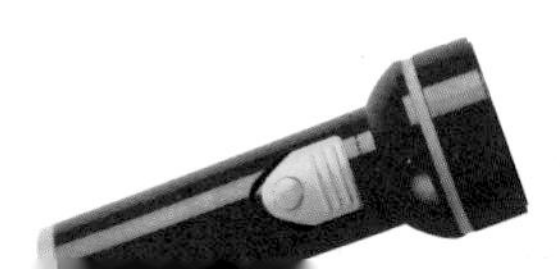